MAY 21

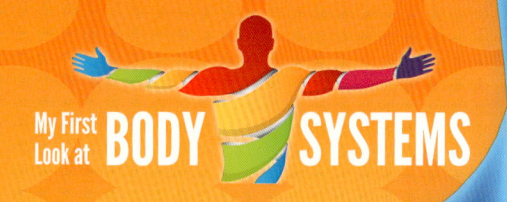

My First Look at BODY SYSTEMS

RESPIRATORY SYSTEM

Becky Noelle
and Priyanka Das

www.av2books.com

Step 1
Go to **www.av2books.com**

Step 2
Enter this unique code

GORMKTRK4

Step 3
Explore your interactive eBook!

RESPIRATORY SYSTEM

My First Look at BODY SYSTEMS

Start!

Your interactive eBook comes with...

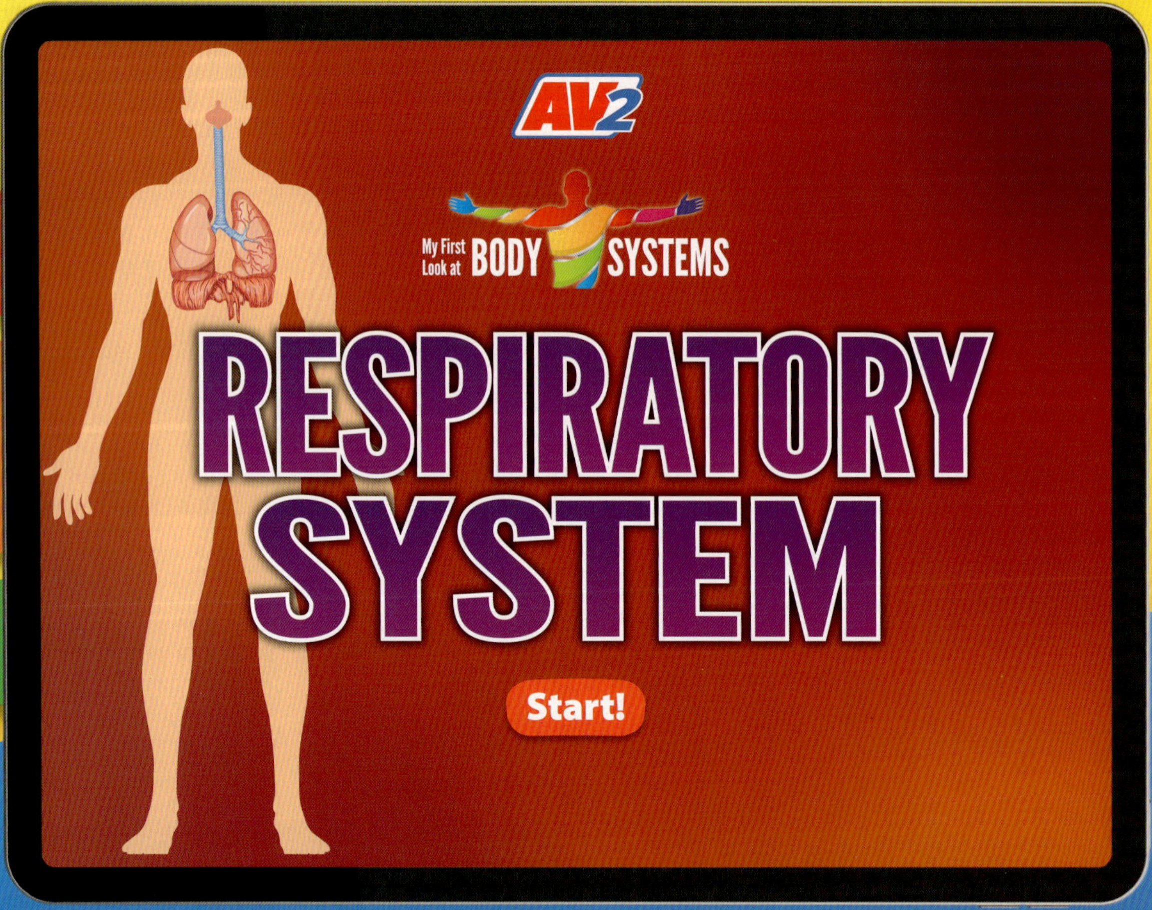

AV2 is optimized for use on any device

Read
Audio
Listen to the entire book read aloud

Videos
Watch informative video clips

Weblinks
Gain additional information for research

Try This!
Complete activities and hands-on experiments

Key Words
Study vocabulary, and complete a matching word activity

Quizzes
Test your knowledge

Slideshows
View images and captions

View new titles and product videos at www.av2books.com

RESPIRATORY SYSTEM

CONTENTS

- 2   AV2 Book Code
- 4   Respiratory System
- 6   All about Lungs
- 8   Parts of the Respiratory System
- 10   Nose and Mouth
- 12   Windpipe
- 14   Inside the Lungs
- 16   Diaphragm
- 18   Staying Healthy
- 20   Career Spotlight
- 22   Respiratory System Quiz
- 24   Key Words

Respiratory System

People use a gas called oxygen for energy. We need it to live. **The respiratory system takes in oxygen from the air.** It also removes a gas called carbon dioxide from the body. This process is called respiration.

The respiratory system is often attacked by viruses. Viruses are very small germs. They can lead to respiratory infections such as colds, the flu, and COVID-19.

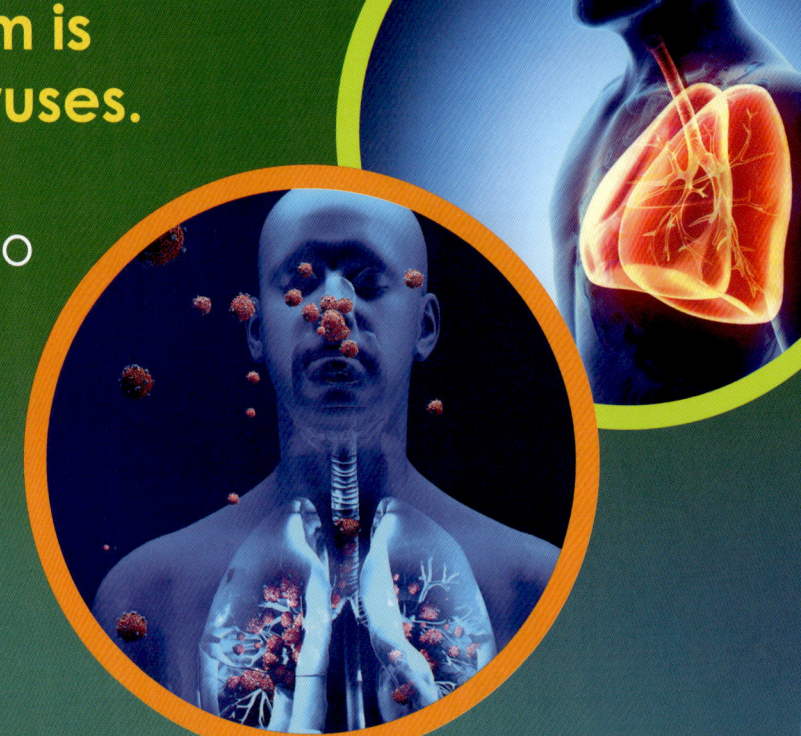

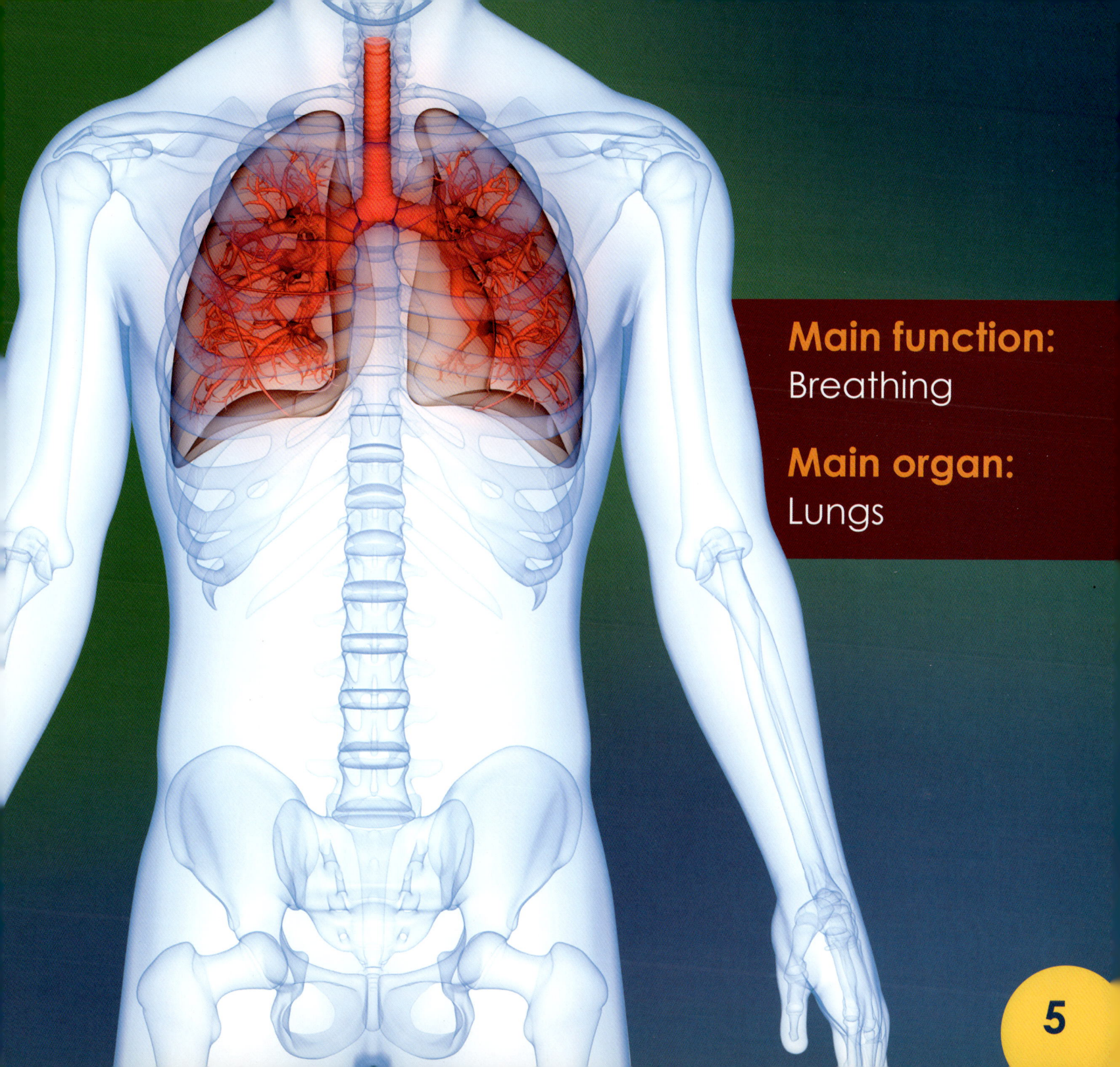

Main function:
Breathing

Main organ:
Lungs

5

All about Lungs

During respiration, people breathe in oxygen and breathe out carbon dioxide. This gas exchange takes place in the lungs. Lungs are very important organs.

The lungs are in the chest. They are so large that they take up most of the space there. **People have two lungs.** The left lung is smaller than the right lung. This makes room for the heart.

People breathe in and out about **20,000 times** every day.

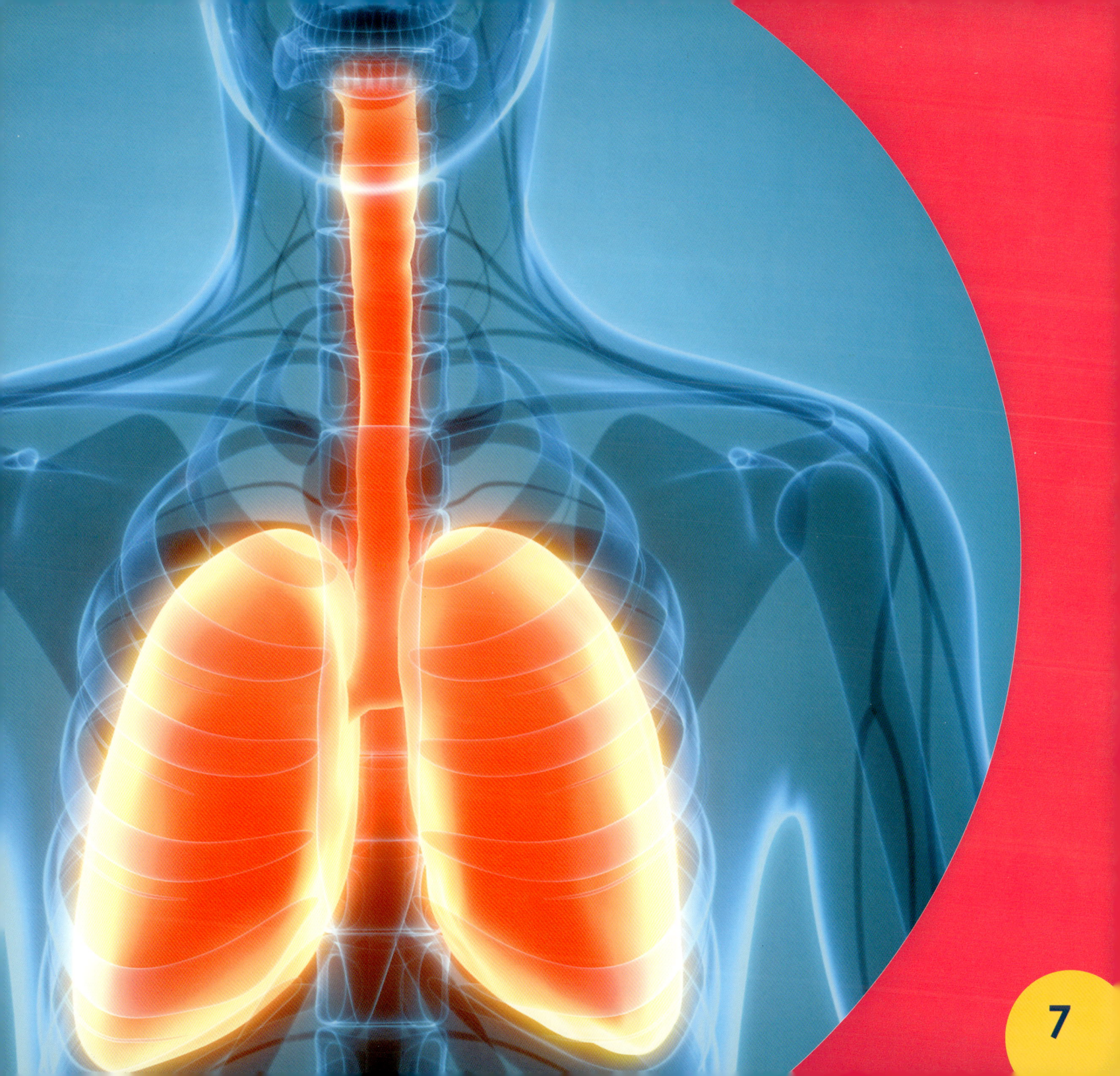

7

Parts of the Respiratory System

When you breathe, air enters your body. This air is full of oxygen. **The air moves down the respiratory tract.** The respiratory tract includes all the body parts involved in the process of respiration.

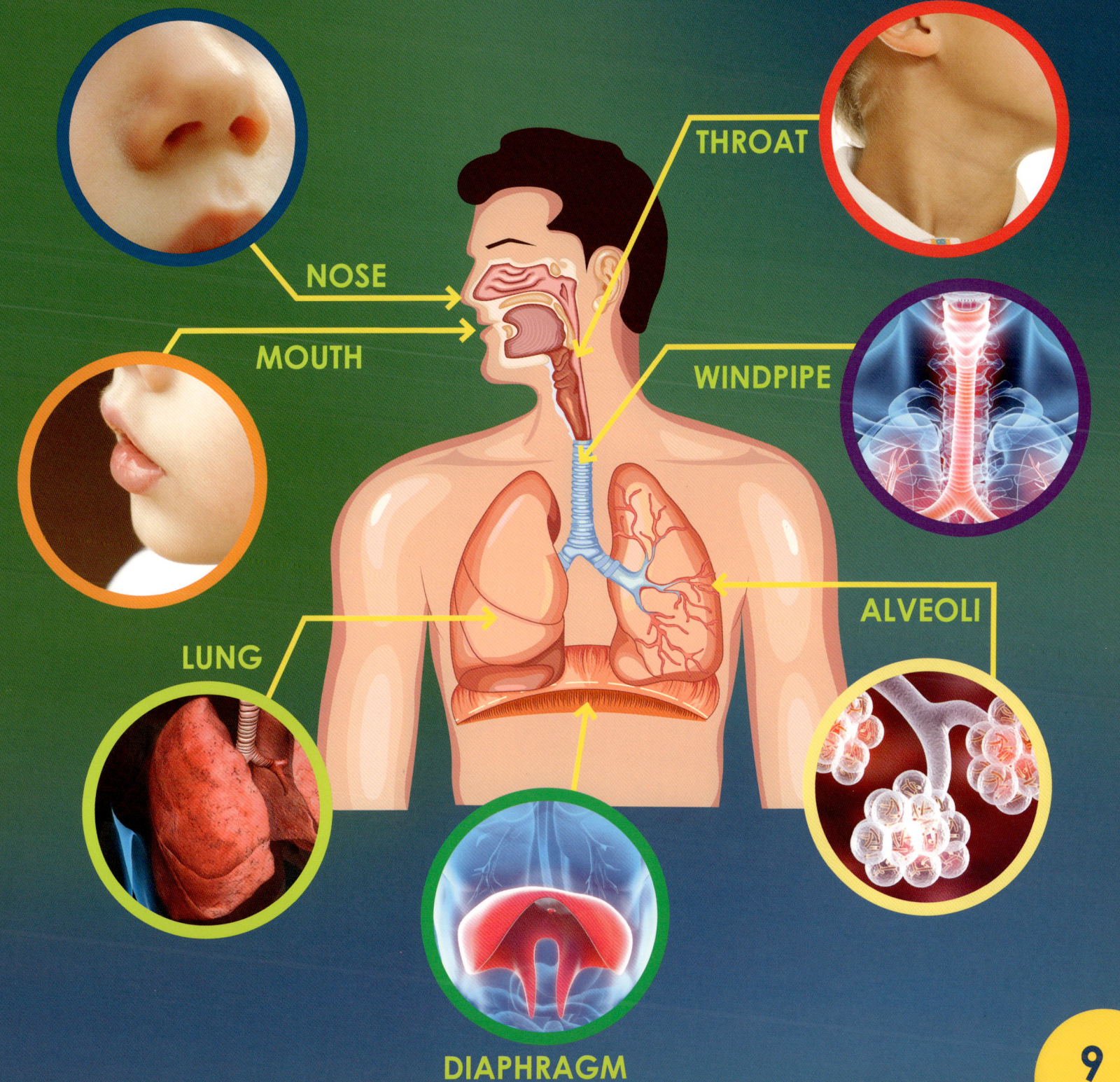

Nose and Mouth

Air enters the body through the nose and mouth. Most breathing is done through the nose. Tiny hairs in the nose remove dust, pollen, and viruses from the air.

People also breathe through their mouths. For example, you might breathe through your mouth when you have a cold. **The nose and mouth lead to the throat.** Air moves from the throat into the windpipe.

Windpipe

The windpipe is a hollow tube. It is protected by rings made of cartilage. These rings make sure the windpipe always stays open.

Air must travel through the windpipe to enter the lungs. **The windpipe's main job is to provide a clear path for air to get in and out of the lungs.**

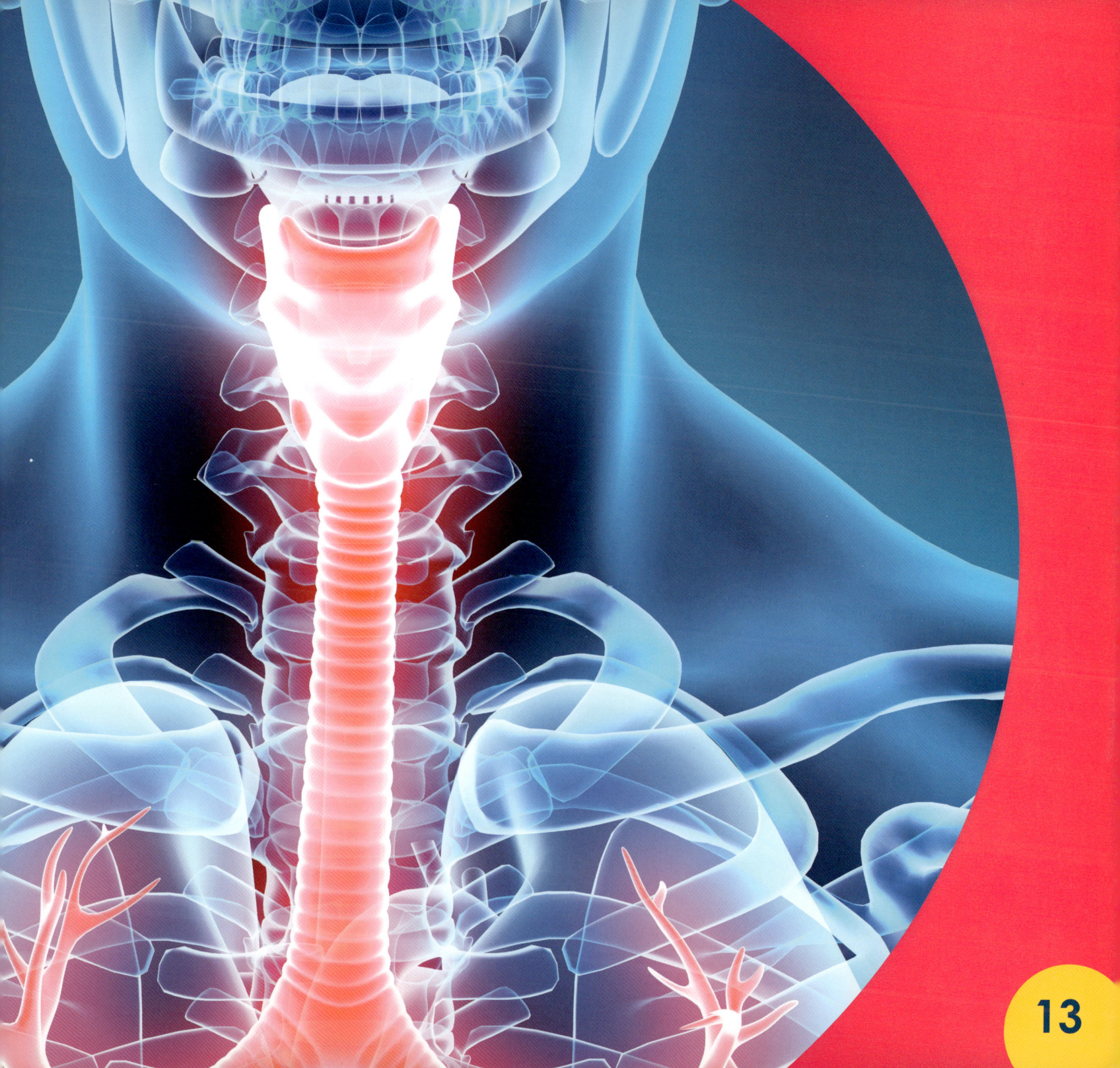
13

Inside the Lungs

The inside of a lung looks like a tree with many branches. The windpipe splits into two tubes called bronchi. These split into smaller tubes known as bronchioles.

At the ends of the bronchioles are tiny air sacs called alveoli. They look like little bubbles. **Alveoli move oxygen from air into the blood.** They also remove carbon dioxide from the blood.

There are about **600 million alveoli** in the lungs.

14

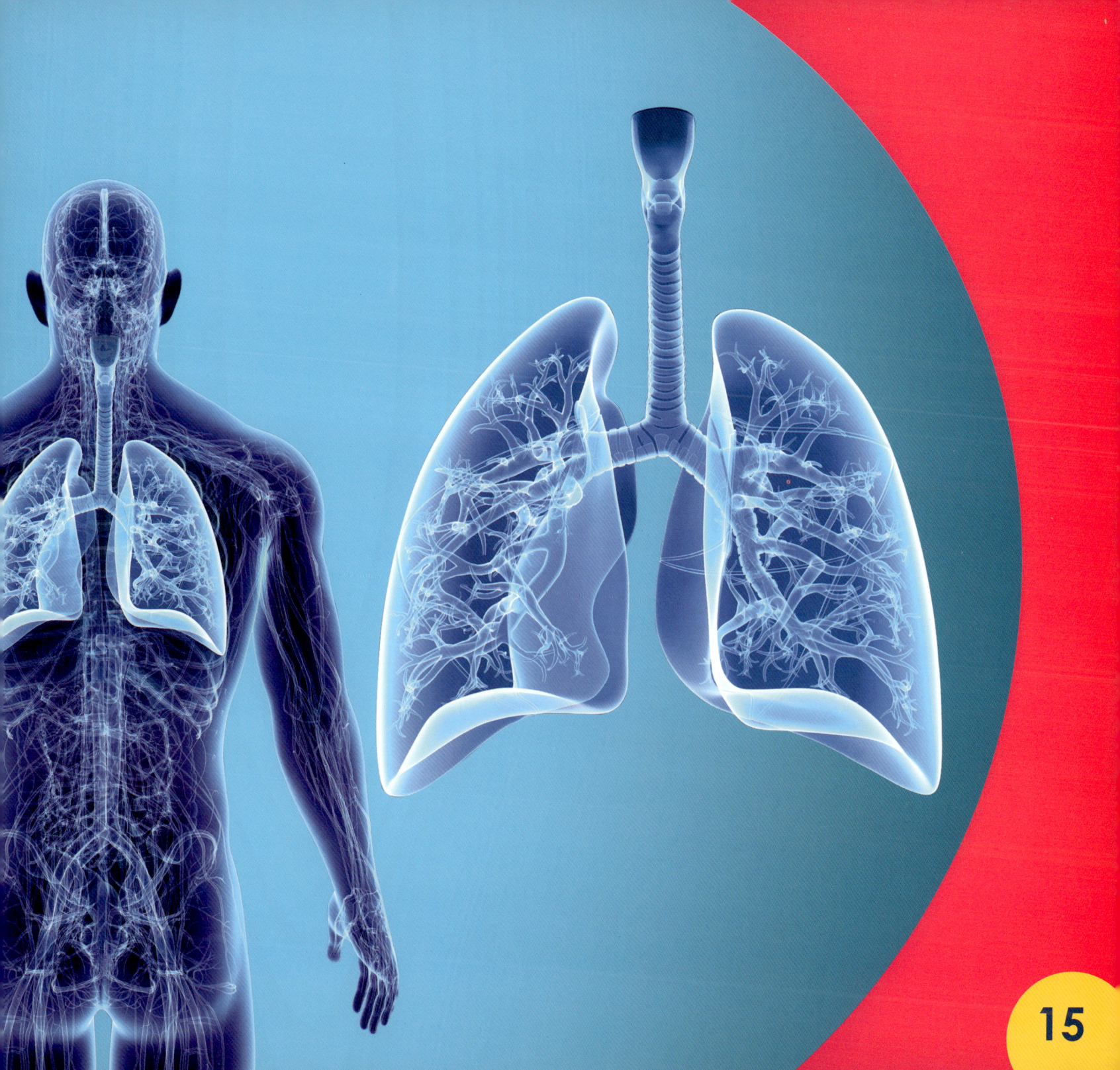

Diaphragm

The diaphragm is a muscle that helps people breathe. It is found below the lungs. The diaphragm is thin and dome-shaped.

When you breathe in, the diaphragm pulls down. **This makes more space for air to move into the lungs.** Then, the diaphragm relaxes. It moves up and pushes air out of the lungs. This is how you breathe out.

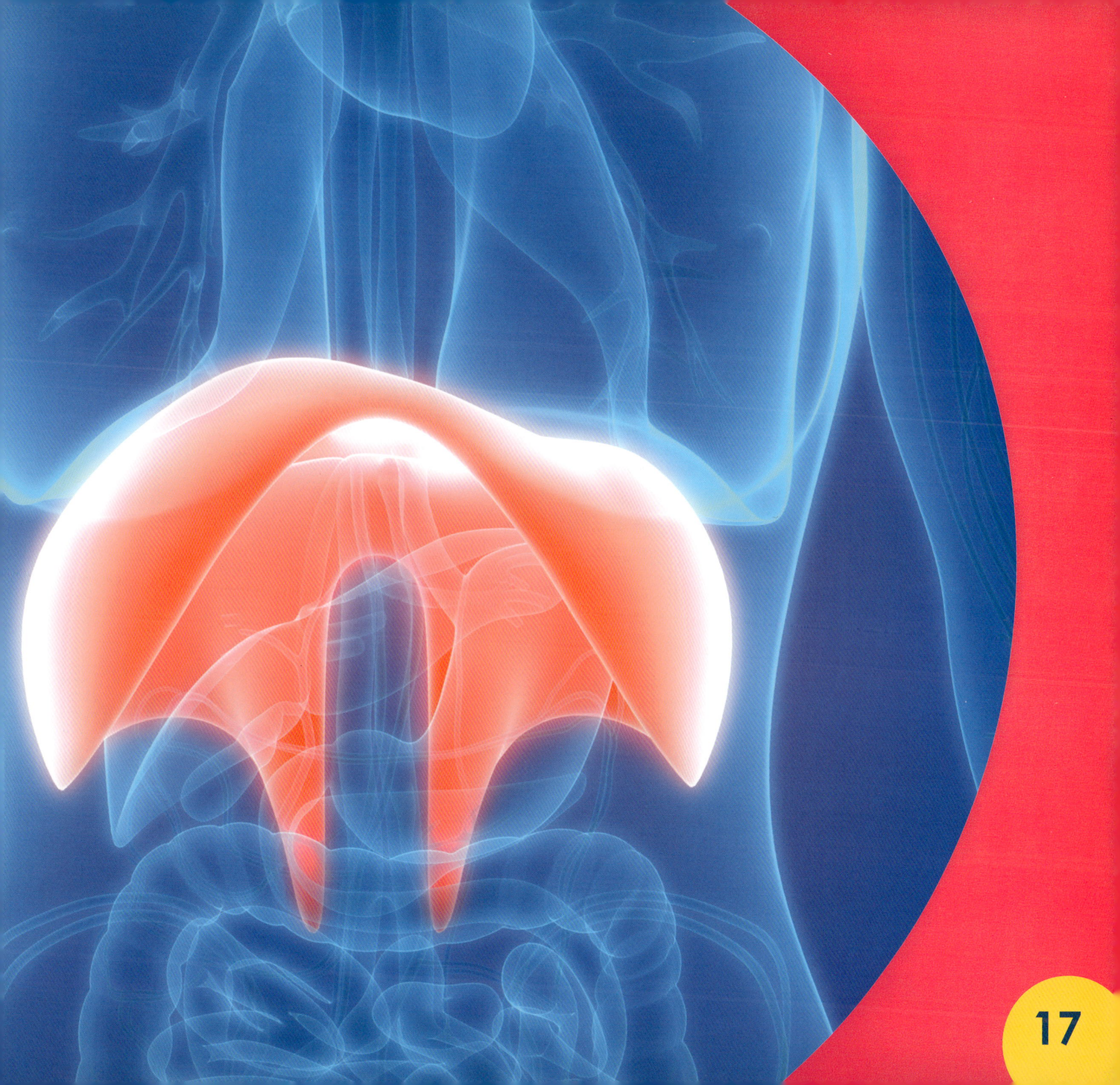

Staying Healthy

Fresh air and exercise help the respiratory system stay healthy. When you exercise, you breathe faster. Your lungs have to work harder to take in more oxygen. This helps the lungs become stronger.

playing

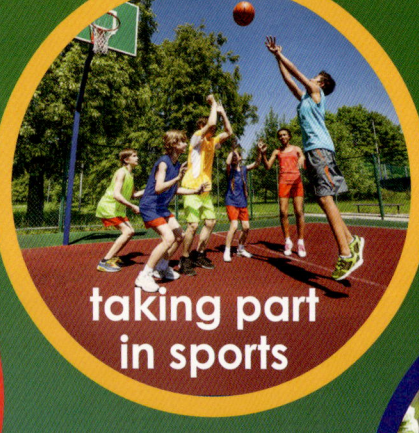
taking part in sports

cycling

running

Dust, pollen, and viruses can hurt your lungs. You can wear a mask when the air is unsafe to help protect your lungs. Washing your hands regularly also helps protect you and others from infections.

wear a mask

wash your hands with soap

cover your mouth when coughing

Career Spotlight

Respiratory therapists help people who have trouble breathing. Sometimes, these people may need a machine called a ventilator. In many hospitals, a respiratory therapist will set up the ventilator. The ventilator breathes for the person. Respiratory therapists also teach people how to take care of their lungs.

OHIO

The Ohio State University is known as one of the best places to learn to become a respiratory therapist.

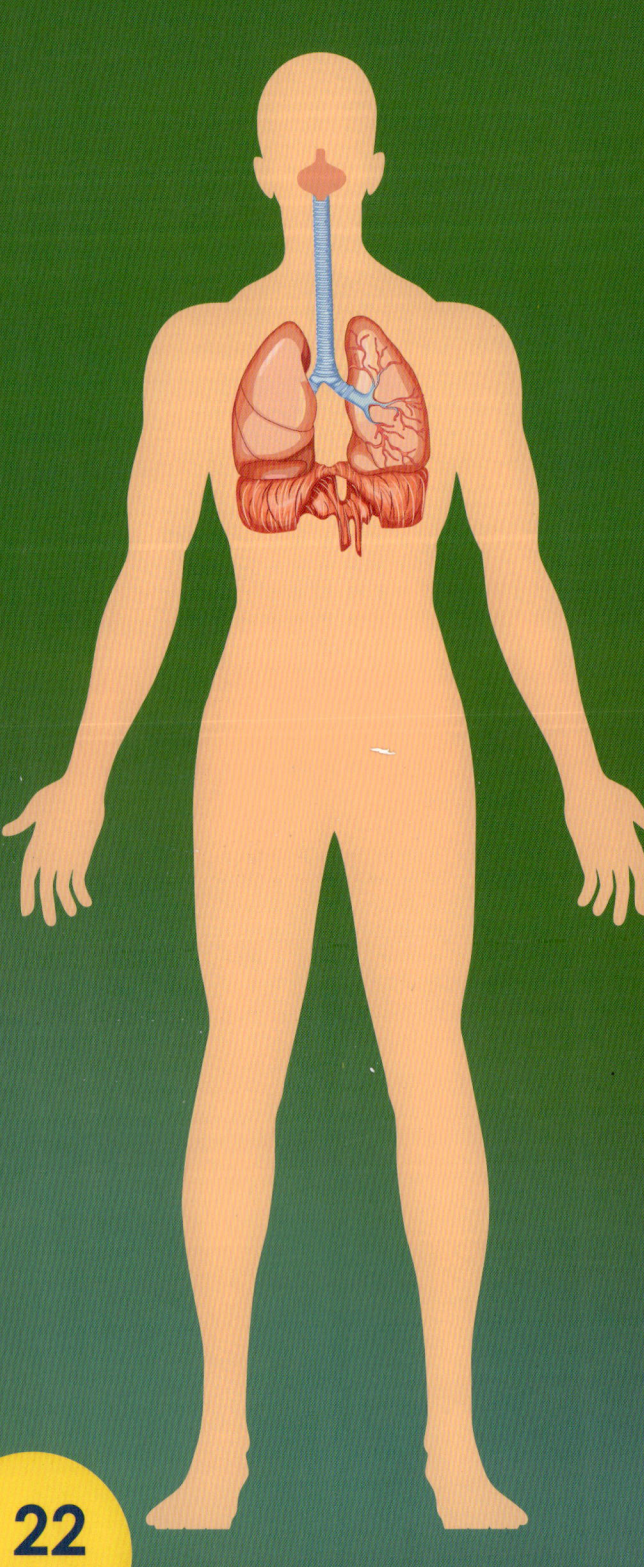

Respiratory System Quiz

Can you name the parts of the respiratory system shown in these pictures?

A] Nose and mouth

B] Windpipe

C] Lungs

D] Diaphragm

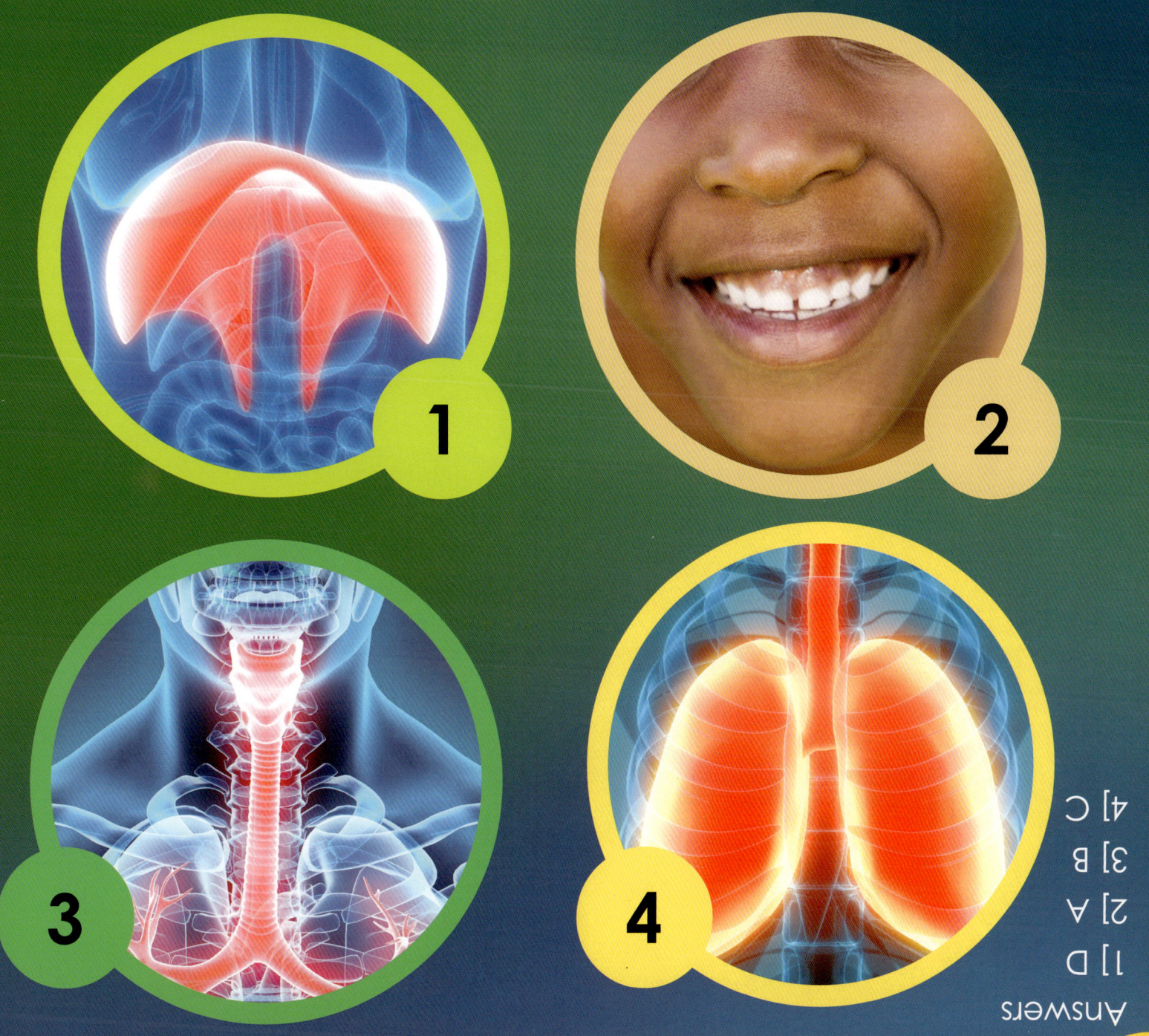

KEY WORDS

Research has shown that as much as 65 percent of all written material published in English is made up of 300 words. These 300 words cannot be taught using pictures or learned by sounding them out. They must be recognized by sight. This book contains 91 common sight words to help young readers improve their reading fluency and comprehension. This book also teaches young readers several important content words, such as proper nouns. These words are paired with pictures to aid in learning and improve understanding.

Page	Sight Words First Appearance
4	a, air, also, and, are, as, by, can, for, from, in, is, it, live, need, often, people, small, such, takes, the, they, this, to, use, very, we
6	about, all, day, every, have, important, large, left, makes, most, of, out, place, right, so, than, that, there, times, two, up
8	down, moves, parts, when, you, your
10	example, into, might, their, through
12	always, get, made, means, must, open, these
14	at, ends, like, little, looks, many, tree, with
16	below, found, helps, how, more, then
18	work
19	hands, others
20	learn, may, one, set, sometimes, state, who, will

Page	Content Words First Appearance
4	body, carbon dioxide, cold, COVID-19, energy, flu, germs, infections, oxygen, process, respiration, respiratory system, viruses
5	breathing, function, lungs, organ
6	chest, heart, space
8	respiratory tract
9	alveoli, diaphragm, mouth, nose, throat, windpipe
10	dust, hairs, pollen
12	cartilage, job, path, rings, tube
14	air sacs, blood, branches, bronchi, bronchioles, bubbles
16	muscle
18	cycling, exercise, running, sports
19	coughing, mask, soap
20	hospitals, machine, Ohio State University, respiratory therapists, ventilator

Published by AV2
14 Penn Plaza, 9th Floor New York, NY 10122
Website: www.av2books.com

Copyright ©2021 AV2
All rights reserved. No part of this publication may be reproduced, stored in a retrieval system, or transmitted in any form or by any means, electronic, mechanical, photocopying, recording, or otherwise, without the prior written permission of the publisher.

Library of Congress Cataloging-in-Publication Data available upon request

ISBN 978-1-7911-1888-4 (hardcover)
ISBN 978-1-7911-1889-1 (softcover)
ISBN 978-1-7911-1890-7 (multi-user eBook)
ISBN 978-1-7911-1891-4 (single-user eBook)

Printed in Guangzhou, China
1 2 3 4 5 6 7 8 9 0 24 23 22 21 20

062020
100919

Project Coordinator: Priyanka Das   Designer: Jean Faye Marie Rodriguez

Every reasonable effort has been made to trace ownership and to obtain permission to reprint copyright material. The publisher would be pleased to have any errors or omissions brought to its attention so that they may be corrected in subsequent printings.

The publisher acknowledges iStock and Shutterstock as the primary image suppliers for this title.